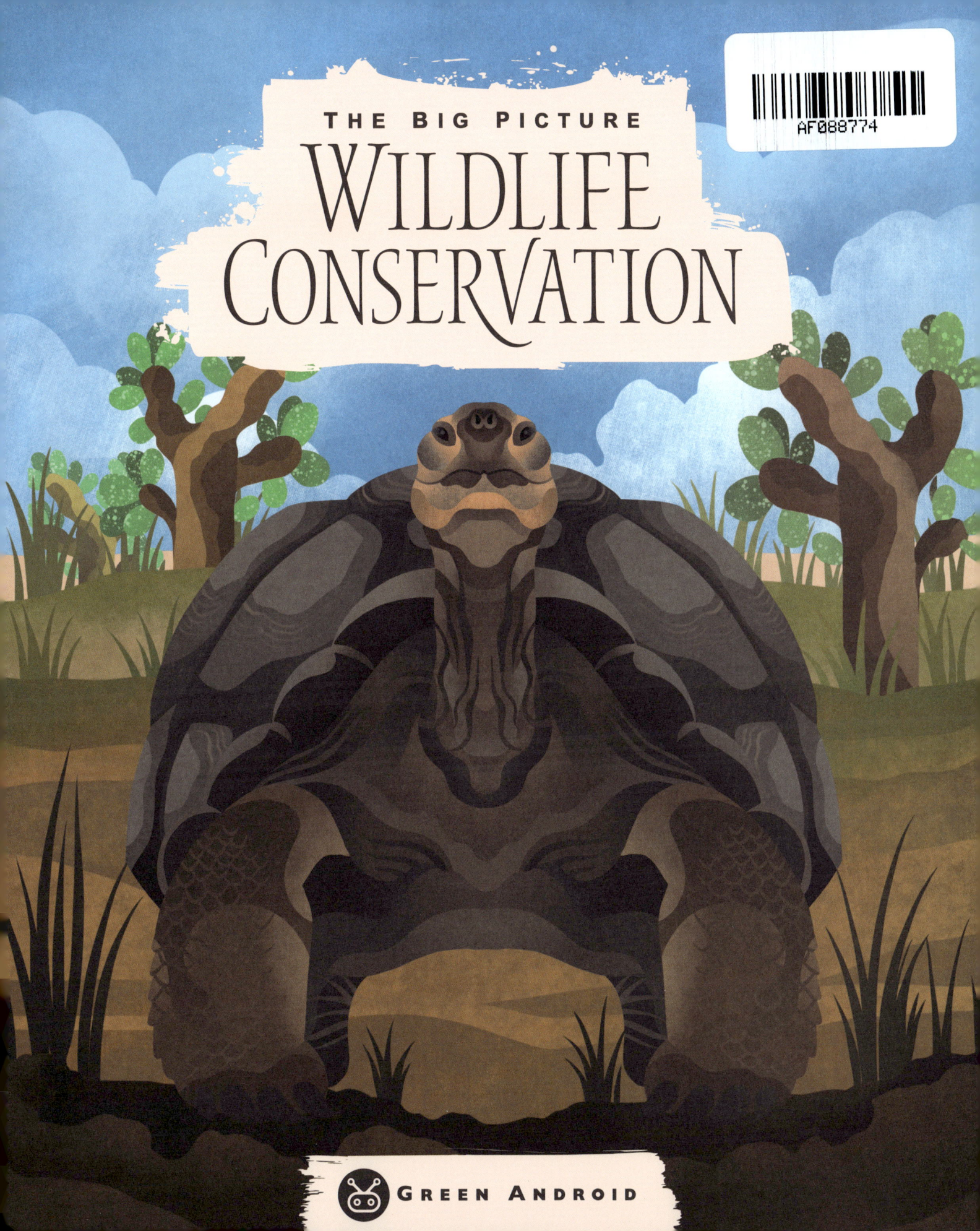

Published by Green Android Limited,
49 Beaumont Court, Upper Clapton Road,
London, E5 8BG

Copyright © Green Android Ltd, 2024

Written by Lyn Coutts
Edited by Katie Wallace and Rachel Cooke
Illustrated by Gilang Ayyoubi Hartanto

All rights reserved. No part of this publication may be reproduced, stored in a retrieval system, or transmitted in any form or by any means, electronic, mechanical, photocopying, recording or otherwise without the prior written permission of the publisher.

A CIP catalogue record for this book
Is available from the British Library

Hardback ISBN 978-1-912188-64-2
Paperback ISBN 978-1-912188-34-5

Date of Manufacture: January 2024
Manufactured by WKT Co., Ltd
Printed in China
9 8 7 6 5 4 3 2 1

Please note that every effort has been made to check the accuracy of the information contained in this book. Green Android Ltd apologise for any unintentional errors or omissions, and would be happy to include revisions to content in subsequent editions of this book.

Contents

A world of wildlife conservation 4

Tigers
6

Giant pandas
8

Bumblebees
10

Eastern gorillas 12

Galápagos giant tortoises 14

Old World vultures 16

Lions
18

Harlequin toads
20

European eels
22

Tasmanian devils 24

Sawfish
26

Kākāpōs
28

Orangutans
30

Sea turtles
32

African elephants 34

Blue whales
36

Rhinoceros
38

Axolotls
40

Lion tamarins
42

Sea otters
44

Conservation in Action 46

Conservation Word Bank 48

A world of wildlife conservation

Today, all around the world, people are working to protect, or conserve, our wildlife. They recognize that many animals and the habitats they live in are in danger.

THERE ARE THOUSANDS OF CONSERVATION ORGANIZATIONS. Some campaign for a particular animal or habitat. Others draw attention to the larger issues that endanger wildlife, such as climate change, farming methods or pollution. Most of these problems are the result of human activity. Campaigners recognize that there is an urgent need to protect nature, not only for people now but also for the generations to come.

"The wildlife and its habitat cannot speak, so we must and we will."
Theodore Roosevelt (1858–1919), US President, conservationist and naturalist

---

In this book we meet twenty different types of animal from around the world that have become the focus of wildlife conservation. They range from tiny insects to huge whales. There are "conservation stars", such as the giant panda and the tiger, but also under-valued species such as the vulture and the eel. What they have in common is that they all appear on the IUCN's Red List of Threatened Species.

There are many, many more species that are in danger than this book can possibly cover. Even this small selection shows the huge range of activity that is taking place around the world to conserve wildlife. You will discover not only the challenges that conservationists face but also their amazing achievements. There are many inspirational success stories that bring hope to endangered wildlife all around the world.

"It's the little things citizens do. That's what will make the difference. My little thing is to plant trees."
Wangari Muta Maathai (1940–2011), Kenyan social and environmental activist who founded the Green Belt Movement for which she won a Nobel Peace Prize

The IUCN's Red List of Threatened Species

The International Union for the Conservation of Nature (IUCN) was established in 1948 to pull together all the available information on the status of the natural world and how to safeguard it. From 1964, the IUCN has produced a Red List of Threatened Species, which categorizes species according to their risk of extinction. Every species on the Red List is constantly being assessed and new species are added. The Red List is updated twice a year.

These are the IUCN's Red List categories:

 Data Deficient: not enough information to assess extinction risk.

 Vulnerable: high risk of extinction.

 Extinct in the Wild: species survives only in captivity.

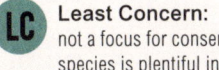 **Least Concern:** not a focus for conservation as species is plentiful in the wild.

 Endangered: very high risk of extinction.

 Extinct: when the last of a species dies.

 Near Threatened: close to qualifying for threatened status.

Critically Endangered: extremely high risk of extinction.

On the Red List of 2022, there were the 150,388 species (animal, plant and fungi), of which over 42,100 are threatened with extinction. In this book, we have just considered some animal species, but all these different life forms are interlinked by the habitats they create and share.

The IUCN Green List of Protected and Conserved Areas

Today there are 261,766 officially listed protected and conserved areas around the globe. They cover 15 per cent of the planet's land surface and 7.4 per cent of the oceans. To acknowledge these activities, the IUCN has created a Green List to work alongside its Red List to highlight good practice and celebrate conservation success stories. It also acknowledges the importance of protecting whole ecosystems as well as individual species.

So far 77 protected areas in 18 different countries have made it to the Green List (December 2022) while another 500 are working toward Green List inclusion. A Green List site must meet the highest level of conservation standards. This includes the strongest protection of nature, with practices that bring about long-term positive impacts on people, nature, wildlife and resources. They also need to have successful responses to threats that range from poaching to climate change. Sites on the Green List are the greenest of green!

Find out more about this, the IUCN Red List and the work of the IUCN at their websites:
iucngreenlist.org • www.iucnredlist.org • www.iucn.org

Tigers

After a century in decline, tiger populations are finally on the rise. This is especially true in India, which is home to three-quarters of the world's tigers.

TIGERS ARE A CHALLENGE TO PROTECT. These top predators range across a huge area (up to 104 square kilometres) to hunt for their prey, which includes wild pigs, water buffalo and antelope. There are six tiger subspecies. Together, they have lost 93 per cent of their historical habitats – rainforest, savannah and grassland – to urban development, industry and farming. As their ranges shrink, the tigers sometimes prey on farm animals. This has brought them into direct conflict with humans.

In one Malaysian forest, patrols and local communities removed 94 per cent of the wire snares set by poachers to trap Malay tigers.

India's Project Tiger

In 2006, there were only 1,411 Bengal tigers in India. In 2023, there are 3,167. This makes India the world's safest place for tigers. Project Tiger declared its first tiger reserve in 1973. Since then, another 53 protected sanctuaries have been added.

Project Tiger aims to build sustainable, healthy populations of Bengal tigers, but also "create a world in which tigers can live". This involves managing their habitats and helping communities live alongside tigers.

● Distribution of all tiger species

Numbers in the wild
3,726–5,578
(All subspecies. IUCN 2022)

Despite this challenge, tiger populations have grown. Why? At the Global Tiger Summit in 2010, 13 Asian countries pledged to double their tiger populations by 2022. Tiger reserves were increased in size and number, and tigers were relocated to them. A wildlife corridor between far eastern Russia and north-eastern China saw Amur (Siberian) tiger numbers triple in just ten years. Across all 13 countries, anti-poaching patrols helped prevent tigers being killed for their fur and body parts.

The South China (Amoy) tiger is the most endangered tiger in the world. It has not been sighted in the wild since the late 1980s, but 100 have been bred in zoos in the Save China's Tigers programme. A handful of these tigers have been transported to a reserve in South Africa, where they relearn their hunting skills. When they are ready, these tigers will be reintroduced into their natural habitat in China, where it is hoped they will breed.

Tigers are still endangered, though. Some regions did not meet their population targets for 2022. However, with ever-stronger conservation programmes in place, there is real hope that the numbers of these majestic creatures will recover in their natural habitats across Asia.

Growing to 3.2 metres from nose to tail, the endangered Bengal (or Indian) tiger is among the largest and most powerful of all cats. Its coat is orange with brown-black stripes.

Giant pandas

The giant panda is living proof that conservation works, although it is still vulnerable. When China protected and replanted vast areas of natural panda habitat, giant panda numbers increased.

IN CHINA IN THE 1950s, the panda's bamboo forests were destroyed for development, carved up by roads, and ploughed for livestock and agriculture. This habitat loss pushed the black and white bear almost to extinction. Confined to small, isolated patches of forest, pandas struggled to find their food (12–38 kilos of bamboo a day) and chances to breed were limited. As the panda population decreased, forest regrowth slowed because fewer pandas were spreading bamboo seeds in their droppings.

In 2020 forest corridors joining 67 separate nature reserves were finalized to make China's Giant Panda National Park. This covers over 26,000 square kilometres. It has led to a huge boost in wild panda numbers. A small number of pandas were bred in captivity and returned to the wild in the park as well. Most of the 600 captive-bred pandas in China are on loan to zoos around the world.

The park is protected by the latest technology. The Digital Panda System has over 600 ground and infrared cameras, plus drones and satellites. These are the park's eyes and ears. At the first hint of a wildfire or other trouble, the system instantly alerts wardens, rangers and the emergency services. It saves the lives of pandas and other animals, and the valuable bamboo forest.

In addition, Chinese researchers have developed facial recognition software to identify individual pandas, so they can monitor a panda's territory, food intake and breeding. It also improves the accuracy of China's panda count. Each count takes up to five years and involves 2,000 people, many trekking on foot. The next count will take place around 2025–2030. Many people are working to protect this shy, elusive and much-loved animal.

Giant Panda National Park (GPNP)

This protected area is home to over 1,600 pandas divided into 18 populations. The GPNP is also home to 641 other animal species and 3,440 plant species.

The vast reserve – three times the size of America's Yellowstone National Park – should give pandas plenty of opportunity to breed. It will also mean that pandas can relocate should the bamboo forest in one area die or be affected by changes in climate.

● Range of giant pandas

Numbers in the wild
1,864
(IUCN 2016)

 **Around 54 per cent of the giant panda's wild and natural habitat in China is now protected.**

The giant panda is seen as one of the "stars" of wildlife conservation because people love this black and white bear. It is even more fitting that it is now a conservation success story.

Bumblebees

Bumblebees are among the best pollinators of plants. Every time a bumblebee moves from one flower to another, it transfers pollen grains that fertilize the plant so it can form fruit and seeds.

WORLDWIDE, THERE ARE 265 SPECIES OF BUMBLEBEE. They are native everywhere except in sub-Saharan Africa, Australia, New Zealand and Antarctica. There is even a bumblebee in the freezing Arctic. Everything from birds to bears relies on the seeds and fruits of bumblebee pollination, and so do humans. We can thank the humble bumblebee for tomatoes, potatoes, peppers, berries, apples, plums and more.

One-third of the food humans eat comes from a crop pollinated by bees. Without bumblebees, our planet would be very different.

Bumblebees help support just about every habitat and wildflower: from woodlands and grassland to our gardens and parks. Despite this, only a small number of species have been studied, and of these 22 are critically endangered, endangered, or vulnerable. Many more have falling populations or are feared extinct.

A quarter of North America's bumblebee species are under threat. The rusty patched bumblebee, for example, was once found in 28 North American states and provinces. Today it is found in just 11 states and one province. It is a key pollinator of cranberries, plums and alfalfa.

Conservation is never simple because an animal is threatened by many different factors. The rusty patched bumblebee has been affected by habitat loss, a reduced variety of plants to feed from, climate change, pesticide use, diseases and pests. As part of the Beenome 100 project, US-government scientists will map the genome of the rusty patched bumblebee and 99 other bumblebees and bees. A genome is the entire set of DNA instructions in a cell. Beenome 100 aims to use this genetic information to help protect these insects from diseases and the negative effects of climate change.

Wildlife Preservation Canada

Since 2014, Wildlife Preservation Canada has been raising captive colonies of the endangered yellow-banded bumblebee. They collect a queen from the wild in the early spring. She is given a nesting site and food in the hope she will produce workers, males, and then queen bees. Queens from these captive disease-free colonies will then be released into suitable wild habitats.

- Range of the yellow-banded bumblebee
- Recent sightings

Numbers in the wild
One-third threatened with extinction
2017 worldwide bumblebee population survey

Like other bumblebees, the rusty patched bumblebee (shown here) uses buzz pollination. While holding the flower, it literally shakes the pollen from the anthers by vibrating its flight muscles at a very high frequency.

Eastern gorillas

For 20 years, until her death in 1985, primatologist Dian Fossey helped slow the decline in gorilla numbers and kickstarted a global conservation campaign.

THERE ARE TWO SUBSPECIES OF EASTERN GORILLA: mountain and Grauer's (eastern lowland). These great apes are found in the Democratic Republic of the Congo (DRC), Uganda and Rwanda, all in central Africa. The status of mountain gorillas improved to endangered in 2018, when their population increased to 1,063. It is hoped that this success can help the Grauer's gorilla, which remains critically endangered.

Mountain gorilla numbers have increased because the entire population live in protected parks. It is a different story for Grauer's gorillas. Their population has decreased by 50 per cent since 1995 because only a quarter are in reserves; the rest are unprotected.

To help Grauer's gorillas, more reserves are being developed and anti-poaching patrols increased. In the DRC's Kahuzi-Biega National Park, rangers are trying to stop illegal logging and charcoal burning which endanger these gorillas. When a charcoal fire gets out of control it burns many square kilometres of the forest in which the gorillas live.

Conservationists here have to work in difficult circumstances. Civil war in central Africa has cost human and gorilla lives. It has destroyed habitats. Many people living in gorilla territories are too poor to buy meat. Instead, some will kill gorillas for bushmeat. Eastern gorillas will remain in danger unless human lives are improved, and this is not easy.

Everyone can help save gorillas by their lifestyle decisions. Buying Forest Stewardship Council (FSC)-certified wood products and sustainable palm oil, and recycling all electronic devices can help reduce logging and mining in gorilla habitats.

Saving the mountain gorilla

Mountain gorillas were saved from extinction because of "extreme conservation". This meant providing safe forests and protecting each gorilla family on a daily basis. There are still dangers for both the gorillas and their guardians from armed rebels and poachers.

The two areas where mountain gorillas live are now national parks. The largest reserve combines the Virunga, Volcanoes and Mgahinga national parks of the DRC, Uganda and Rwanda. The other area is the Bwindi Impenetrable National Park in Uganda.

● Distribution of Grauer's gorilla
● Range of mountain gorilla

Numbers in the wild
Less than 5,000
(Both subspecies. IUCN 2018)

The gorillas in the Virunga parks have the highest level of protection of any animal. Over 770 rangers patrol the parks' 1,004 mountain gorillas.

Mountain gorillas live at high altitudes where temperatures can drop to below freezing. As human settlements force gorillas to higher elevations, they have to survive dangerous conditions for longer.

Galápagos giant tortoises

The Galápagos Islands are off the Pacific Ocean coast of South America. This unique marine reserve is home to giant tortoises that can live to 100 years old.

SIX OF THE 12 SURVIVING SUBSPECIES of Galápagos tortoise are critically endangered. Each subspecies has evolved to the habitat of its birth island, adapting to the diet it provides. These large reptiles forage for eight hours a day eating cacti, grasses, lichens and some fruits. Giant tortoises spread seeds, trample bushes and thin vegetation, making them the world's slowest-moving (0.3 kph) ecosystem engineers.

Rewilding giant tortoises

The Galápagos Islands' giant tortoise captive breeding and rewilding programme has been hugely successful. About 9,000 juveniles have been released to their natural homes.

In 1971, there were just three male and 12 female Española giant tortoises on Española Island. Over the next 33 years, 1,200 juveniles were raised, released and tagged, so they can be tracked via satellite. The giant tortoise population on Española is now self-sustaining, so the breeding programme has stopped.

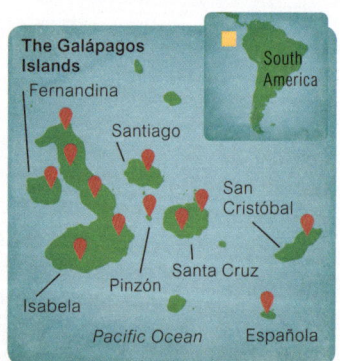

Distribution of Galápagos giant tortoises

Numbers in the wild
24,000–27,000
(All subspecies. IUCN 2017)

With no natural predators on their islands, giant tortoises once numbered over 200,000. But this was reduced to 15,000 by the 1970s after settlers introduced livestock, dogs, rats and cats to the islands. Livestock squashed tortoise nests under their hooves, while other introduced animals raided nests for their eggs or ate the hatchlings. Giant tortoises were also hunted by seafarers and settlers for their meat and shell.

The improvement in giant tortoise numbers began when captive breeding programmes started in 1965. This requires removing an egg from a clutch, incubating it, and rearing the hatchling for six to eight years. The juvenile is then returned to its native island. The captive breeding and rewilding programmes have saved some subspecies from extinction.

With this in mind, the story of the Fernandina giant tortoise is awaiting a conclusion. A single specimen of this fantastic giant tortoise subspecies with its flared shell was recorded in 1906 but not seen again until in 2019. Then a lone female, named Fernanda by scientists, was found on Fernandina Island feeding on the few plants that grow among the rocks of its active volcano. Scientists are now hunting urgently for a mate for Fernanda so that this subspecies can survive.

Charles Darwin's study of Galápagos giant tortoises contributed to his theory of evolution, *The Origin of Species*.

The island of Santa Cruz is home is to Western and Eastern giant tortoises. The Western species (illustrated here) has a strongly domed shell, grows to over 1.2 metres in length and weighs 408 kilos.

Old World vultures

These vultures, native to Asia, Africa and Europe, are the world's most successful scavengers and the highest-flying raptors of all. They are also under threat, but innovative conservation is paying off.

OF THE 16 OLD WORLD VULTURE SPECIES, the IUCN class 11 as vulnerable, endangered or critically endangered. People are often scared by vultures and their reputation as scavengers, but they play a vital role in keeping ecosystems working. By eating the carcasses of dead animals, they clean away diseases that can be deadly to wildlife and humans. For example, they remove the lethal rabies virus which can be spread via feral dogs to humans.

Over 30 years, from 1990 to 2020, African vulture populations fell by 80 per cent. This followed on from a similar crisis in India. The greatest threat to vultures is ingesting toxic chemicals from the dead animals they eat.

Some of these are accidental poisonings, from drugs injected by vets into livestock, but others are deliberate. Some African farmers plant poisons to kill the cheetahs and lions that attack their livestock. These poisons also kill the vultures. Poachers will also deliberately poison vultures to stop them circling over a dead animal's body before they feed. The poachers realize that this sight may alert rangers that they are active in the area.

To save the lives of endangered vultures, some are fitted with GPS trackers. Weighing just 71 grams, the trackers transmit data hourly. If a vulture has not moved or its body temperature is low, a field team heads to its location. They will rescue the injured, ill or poisoned vulture. They will also dispose of a poisoned carcass. By removing it, the team prevents other vultures from being poisoned. GPS devices and field teams have reduced vulture deaths by a third.

Vulture Safe Zones (VSZ)

In India, Nepal and Bangladesh there are 12 VSZ "restaurants" where wild vultures can scavenge poison-free cattle. Cows that can't produce milk (cows are sacred to Hindus, so it is illegal to slaughter them) die naturally in rescue centres and become safe food for the vulture.

South Africa's first VSZ is in Tswalu Kalahari. This safe zone aims to conserve the critically endangered white-backed vulture. This species faces not only poisoning, but habitat loss, electrocution from power lines and illegal poaching.

- Distribution of Old World vultures
- Safe zones

Numbers in the wild
Up to 1,519,668 mature individuals
(15 species. IUCN 2016, 2020, 2021)

Old World vultures are nature's clean-up crew. A flock can strip an antelope in 20 minutes.

The critically endangered white-backed vulture (left) is found in southern Africa only. The endangered lappet-faced vulture (right), with its distinctive skin folds or lappets, inhabits southern Africa and the Arabian Peninsula.

Lions

Conservation is slowly bringing the lion's roar back to Africa and to India's forests and savannahs. But while this royal beast remains Red Listed as endangered, it must be protected.

LIONS ONCE INHABITED AFRICA, southwest Europe and parts of Asia. Today, the two subspecies of *Panthera leo* are found in less than one-tenth of their historic range. Populations have increased in India and some African countries. But overall numbers in Africa fell by 43 per cent in just 20 years. These apex predators prey on zebras, wildebeests and other herbivores. Lions help maintain the balance of a habitat, ensuring that forests, grasslands and shrublands where they live are not overgrazed.

Asiatic lion sanctuary

In India, the Asiatic lion population is confined to the Gir Forest National Park in Gujarat. This sanctuary is home to about 550 lions – a huge increase on the 20 that remained 110 years ago when this was a hunting park.

Though highly protected and patrolled, this single population is concentrated in such a small area, it could be wiped out by disease or a forest fire. To prevent this, rangers use an electronic monitoring system called SMART. At the first sign of any danger, rangers and emergency services can respond.

● Distribution of African lion
◉ Range of Asiatic lion

Numbers in the wild
23,000–39,000
(Both subspecies. IUCN 2014)

Warriors who once proved their manhood by killing lions are now guardians: they protect the lions for the benefit of their community.

Humans are the main cause of lion deaths, especially in Africa. Lions that come too near settlements or livestock are shot or poisoned, in what is known as "revenge killings". Only half of Africa's lions live in protected areas; the rest are forced into close and often fatal contact with humans.

People need to learn to live alongside lions and respect their need for space. To reduce revenge killings, there are charities that install fencing to protect livestock from attack. In areas where farmers are compensated for livestock taken by lions, fewer lions are shot.

There are also community education programmes that highlight the economic benefit that a well-managed lion reserve can bring to local people, for example through tourism. The message conservationists want to spread is: lions are worth more alive than dead.

The Samburu warriors of Kenya, Africa, monitor the lions in their area, and inform herders and the community of the lions' location in order to reduce contact. In this way, the warriors maintain their traditional role of protecting their people at the same time as conserving the lion population in the region.

Popularly king of the beasts, African lions (illustrated here) are larger than Asiatic lions. They also have thick, long manes, whilst the Indian lions have short, sparse manes.

Harlequin toads

Measuring only 3.8 cm from their pointy snout to the end of their colourful body, harlequin toads are important predators of insects in Central American rainforests. They indicate high quality water too.

THERE ARE 100 SPECIES OF HARLEQUIN TOADS. Because of their jewel-like colours and smooth skin (toads usually have dull brown, warty skin), harlequin toads have been mistaken for frogs. But like all toads, the glands on their skin secrete toxins. Harlequin toads live in mountain areas of tropical rainforests. Once abundant across 11 Central and South American countries, a dramatic fall in their numbers was first recorded in 1988. Scientists are working hard to save these remarkable toads.

Scientists have successfully bred some harlequin toads in captivity. They aim to reintroduce these to the wild.

Restoring harlequin toads

In 2021, 40 organizations from 13 countries agreed to share their expertise in an effort to prevent more harlequin toads becoming extinct in the face of the chytrid disease that, at present, can't be beaten.

They are developing ways to lessen chytrid's impact and discovering why some species are more vulnerable to the disease than others. Field teams are searching for the toads in the hope that they will find new, uninfected populations of this secretive amphibian.

Habitats of all species of harlequin toad

Numbers in the wild
62 critically endangered species
(from 100 known species, IUCN)

As ever, there are many reasons for the toad's decline: habitat loss, climate change and water pollution. Harlequin toad tadpoles became prey for introduced fish species, and juveniles and adults were captured for the exotic pet trade.

However, the biggest cause for their decline was disease caused by a chytrid fungus. It is infectious and fatal for most amphibians, including harlequin toads. Chytrid prevents an amphibian's skin taking in sufficient quantities of salt and water. The animal will begin to die after 21 days. There is no cure. Scientists believe it has made 70 per cent of harlequin toad species critically endangered, and as many as four species extinct.

One hope for saving harlequin toad species includes captive breeding programmes. In the case of the very rare variable harlequin toad, scientists recreated the exact conditions – lighting, temperature, water flow, rocky stream bed and the tropical algae – that six black and yellow toads needed for breeding and for their tadpoles to survive. The programme took place at the University of Manchester in the UK – an ocean away from the toad's natural home. It was the first time this toad had bred outside of its country of origin.

The variable harlequin toad can have neon green, yellow, orange or pink spots and streaks against a black background. In the international pet trade, it might be sold under the name "clown frog". It breeds in fast-moving rainforest streams.

European eels

European eels are snakelike fish that undertake an extraordinary migration. Once abundant, the species is now critically endangered. Action is underway in Europe to save this remarkable creature.

EUROPEAN EELS START LIFE IN THE SARGASSO SEA in the Atlantic Ocean. Here, millions of leaf-shaped larvae hatch and ride ocean currents eastwards for 6,500 kilometres. On reaching the UK, mainland Europe and North Africa, the larvae change into 5 cm-long glass eels, moving into freshwater river systems. They head upstream, changing again into darker elver and then yellow eels. They hunt at night and hide during the day. After at least 20 years, they change again into silver eels, ready to reproduce. These mature eels head downstream and back to the Sargasso Sea to mate and lay their eggs. The eels die but the cycle begins again.

In Europe, the turbines of some hydroelectric power stations are turned off at times of eel migration.

Opening the River Thames

European eels can spend 20 to 30 years of their life in the UK's River Thames, feeding and growing. But along this 346-kilometre river and its tributaries, there are over 2,400 artificial barriers that could prevent eels migrating up or down river. The Thames Estuary Trust and its 22 partners are restoring the eel's migratory pathways.

Over 1,000 volunteers are involved. To turn a barrier into an eel pass sometimes just requires the installation of a ramp lined with a brush-like surface that eels can climb.

- Range of European eel
- → Eastward migration
- → Westward migration

Numbers in the wild
95 per cent decline
(In past 45 years. IUCN 2014)

Since the 1980s, European eel numbers have fallen by 95 per cent. This decrease has multiple causes. Loss of wetlands and the installation of weirs, dams, pumping stations, flood defences and more, prevent juvenile eels getting into fresh water where they can mature. River pollution affects the reproductive ability of eels, while many eels are infected by damaging parasites.

Any one of these factors is enough to prevent tens of thousands of eels failing to complete their life cycle. In addition, people all around the world eat eels. For many, they are a great delicacy.

There has been a ban on the export of live European eels since 2010. But some think around 350 million eels are smuggled from Europe to the Far East each year. This is partly why the IUCN lists the European eel as critically endangered.

All around Europe different conservation organisations are finding ways to help eels make their migration. This includes persuading power companies to turn off their hydroelectric turbines at times so eels can pass safely through a dam. Governments are tightening the policing of the illegal fishing and export of live eels.

European eels can live for up to 85 years. In their yellow and silver stages, eels eat small fish, frogs, insect larvae, and dead and rotting creatures. Eels are prey for eagles, osprey, bass and snapping turtles. Many humans like to eat them too.

Tasmanian devils

Tasmanian devils are the world's largest living meat-eating marsupials. These fierce, unique animals are endangered, dying of a contagious disease. Scientists are working on a vaccine and a cure.

THE TASMANIAN DEVIL IS FOUND ON THE ISLAND STATE OF TASMANIA, Australia. About the size of a small dog, it has a bloodcurdling screech. The bite force of its jaws is the greatest of any predatory land mammal. It can bite through thick wire! This carnivore hunts small prey or scavenges on dead animals, helping clean and clear its habitat. The Tasmanian devil eats all of an animal. It leaves nothing, not even the bones.

Healthy Tasmanian devils that have been relocated to zoos and protected reserves are known as "insurance populations".

Numbers in the wild
10,000–25,000 mature individuals
(IUCN 2008)

Devils in Tasmania survived centuries of historic climate changes and being preyed on by other animals or hunted by humans. Their population numbered 150,000 in the mid-1990s but has fallen by over 80 per cent. The dramatic decline is caused by a deadly contagious disease that appeared in 1996. The devils pass on the disease when they bite each other during mating and dominance fights.

Since 2013, about 500 healthy Tasmanian devils have been relocated to zoos and protected reserves in Australia and around the world. These are "insurance populations" in case the disease causes devils to become extinct in the wild. One insurance population of 28 devils, moved to Maria Island off the southeast coast of Tasmania, grew to 90 within four years.

Return to the mainland

In 2020, 28 Tasmanian devils were moved from Tasmania to mainland Australia. These are the first wild devils on mainland Australia for 3,000 years. The mainland's devil population was most probably wiped out by dingoes, Australian wild dogs, around 3,000 years ago.

The Tasmanian devils' new home is at Aussie Ark in New South Wales' Barrington Tops. This 4 square kilometre sanctuary is fenced and each devil is monitored. In 2023, three pups (or joeys) were born. The ultimate goal of this project is to rewild Australia with Tasmanian devils.

● Current range of Tasmanian devils
● Rewilded site

In addition, scientists are developing a vaccine for the disease that will protect healthy devils. They hope to vaccinate the animals via an edible bait from a device which only devils will be able to access. In 2023, a cell treatment was developed that can cure Tasmanian devils infected with the disease. Both vaccine and cell treatment are very good news for the future of the Tasmanian devil.

Tasmanian devils stand about 30 cm tall and weigh up to 13 kilos. A white stripe runs across their brown-black chests. Devils have very sharp senses of smell and hearing. Like other marsupials, the pups (or joeys) develop inside their mother's pouch.

Sawfish

Sawfish are large rays. They are named after their "saw", the toothed rostrum that extends from the front of their body. These saws often feature in myths as symbols of protection, containing powerful spirits.

THERE ARE FIVE SPECIES OF SAWFISH found around the world. Of these, the IUCN list four as critically endangered and one as endangered. Sawfish can grow to 7 metres, including the rostrum. This is around a quarter of the ray's length. The rostrum has sense organs that help the sawfish locate, slash and stun its prey, such as small fish. The rostrum can easily become tangled in fishing nets. Once common in the waters of 78 countries, now sawfish are only found in the waters of 36.

Sawfish are reliably found in places where they are fully protected and fishing is policed, such as the US and Australia.

Largetooth sawfish rescue

In Australia's Northern Territory, rangers have been saving largetooth sawfish since 2012. They rescue the sawfish stranded in billabongs (ox-bow lakes), waterholes and creeks that are drying up. In just one event, 39 sawfish pups were rescued from an 18-centimetre-deep muddy waterhole.

The rangers tag the sawfish and drive them to permanent water in the Daly River. The sawfish may stay here for five years until they migrate to the sea.

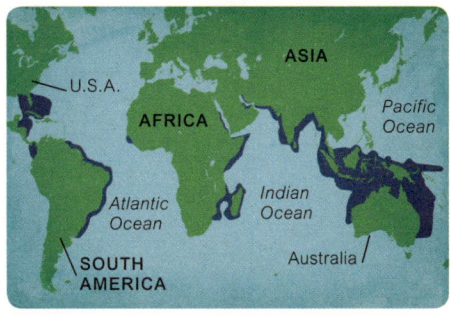

● Common range of all species of sawfish

Numbers in the wild
50–95 per cent decline
(Since records began, all species. Olive Ridley Project estimate, 2021)

Around 90 per cent of the global population of sawfish has been lost. They were overfished for centuries for their meat, skin, body parts and rostrum. The rostrum features in many cultures' myths. It was thought to hold powerful spirits that protected its owner. Later it was also in high demand as a decorative curiosity.

Despite an official worldwide ban, the trade in sawfish thrives. Local fishing communities need to be encouraged to preserve their sawfish populations. Sawfish are reliably found in Northern Australia and Florida, USA, where they are fully protected and fishing is policed.

Sawfish generally live in brackish coastal water and estuaries. Some species move upriver to fresh water. Sawfish can live for 30 years or more. The female gives birth to live young in shallow water, such as mangrove habitats. These provide plenty of food while tangled roots protect the young from predators.

Mangrove habitats are threatened by pollution (especially herbicide run-off from land), oil spills, sand dredging, development, logging and sea level rise. Where mangrove habitats are protected around the world, sawfish numbers are slowly improving.

The largetooth sawfish is critically endangered. Its robust rostrum has 12 to 24 equally-spaced teeth on each side of the saw. Like all sawfish, its main diet is fish, crabs and shrimps. It lives both in freshwater river and brackish coastal habitats.

Kākāpōs

The kākāpō is the world's only flightless parrot. Found in New Zealand, 50 years ago, kākāpōs were thought to be extinct, but intensive conservation is bringing back these fascinating birds.

KĀKĀPŌS ARE UNIQUE TO NEW ZEALAND. They are the world's heaviest and only flightless parrot. Until the arrival of people on the islands, kākāpōs had no natural predators. They lost their ability to fly over thousands of years. Instead they climb trees and descend by parachuting down on their small wings. Kākāpōs can live for over 100 years.

Since Māoris arrived in New Zealand in the 1300s, kākāpōs have been hunted for their meat and feathers. Then, in the 1840s, the British colonized the islands, bringing with them cats, rats, ferrets and stoats. These invasive species preyed on the flightless kākāpō. Their numbers dropped rapidly.

By the 1970s, it was feared the bird was extinct. Then small populations were found in remote parts of New Zealand. In 1995, the Kākāpō Recovery Programme was set up, which still operates today. It relocated all the kākāpōs on the mainland to five small islands that had been cleared of all predators. Every kākāpō is microchipped and fitted with a radio transmitter so it can be tracked. Even their nests are monitored by cameras and sensors. The birds are also given an annual health check and treated for illness, such as a recent fungal infection.

If a kākāpō mother needs help incubating an egg, kākāpō rangers remove the egg. While the real egg is being artificially incubated, a "smart egg" is put into the nest. The dummy egg mimics the temperature and sound of a real chick. This means the mother kākāpō is ready to look after a newly-hatched chick when it is returned. Kākāpōs remain critically endangered on the IUCN Red List, but slowly their numbers are growing.

Return of the kākāpō

Until 2023, the kākāpō population had been kept safe and isolated on offshore islands. But these became overcrowded. In July 2023, four male kākāpōs were moved to Sanctuary Mountain Maungatautari on New Zealand's North Island.

This 34-square-kilometre park is fenced to keep out predators, so it is ideal for the kākāpō's historic return to the mainland. The plan is for more males to be brought to the sanctuary. Once these are settled, females will join them.

- Current distribution
- Rewilding sites

Numbers in the wild
248
(Dept of Conservation, New Zealand, 2023)

Three-dimensional printed "smart eggs" that mimic the sounds of an incubating egg are placed in kākāpō nests.

Kākāpōs are lek breeders. The males gather at a site and call out for the females. A female chooses her mate by the quality of his call. They do not mate for life. The female builds the nest, incubates the eggs and raises between one and four chicks on her own.

Orangutans

All three species of orangutan are critically endangered. There are many challenges in protecting them, but people are fighting to save this great ape, one of the closest relatives of humans.

ORANGUTANS WERE ONCE WIDESPREAD THROUGHOUT SOUTHEAST ASIA. Today, they are only found on the Malaysian and Indonesian islands of Sumatra and Borneo. As "gardeners of the forest", orangutans are responsible for seed dispersal. They spread the seeds of the fruits they eat in their droppings. Some tree species would disappear if it weren't for orangutans. Orangutans keep forest ecosystems in balance.

Orangutans rescued from the pet trade are taught to survive in the wild so they can be released into national parks.

Orangutans are especially vulnerable. They are slow to reproduce. A female will give birth to just one young every eight to ten years. They move slowly, too, making them easy to hunt and catch. It has been illegal to hunt or trade orangutans since 1931. Poaching continues because the financial gain is high and the risk of imprisonment or fines is low. Up to 3,000 orangutans a year end up as trophy pets or are killed for bushmeat, even in protected parks. Of the 41 reserves in Indonesia, 34 are illegally logged or mined and workers kill orangutans for their meat.

Cutting down or damaging forests for agriculture, mining and development take a high toll on orangutan numbers. Where once there were pristine forests, there are now millions of square kilometres of palm oil trees. Yet encouraging research shows orangutans are more adaptable than once thought. They can thrive at higher altitudes and in forests that are selectively logged and managed well. A way forward, conservationists believe, is to guarantee that palm oil plantations are crossed by natural corridors so that orangutans can move between forests to feed and breed. It is vital too that local communities are engaged to help protect these gentle animals.

Protecting the Tapanuli

Tapanuli are the rarest species of orangutan. There are fewer than 800 individuals and they are found only in Sumatra's Batang Toru Ecosystem. There are few government controls here to protect the orangutan's habitat. Instead, a charity called the Wildlife Trade Monitoring Group carries out patrols and reports land clearance. They educate and work with the local community. Together, they have recently delayed the building of a huge dam in the area.

- Distribution of all orangutans
- The Batang Toru Ecosystem

Numbers in the wild
118,862
(All species. IUCN 2016–2020)

Like all orangutans, the Sumatran orangutan spends 90 per cent of its time up in the trees. They forage there for their food, mainly fruit. They build nests from the leaves as places to rest.

Sea turtles

Decades of conservation work have led to increases in sea turtle nests for some of the seven sea turtle species. But, despite these successes, turtles are still in danger. There is work to be done.

THESE REPTILES SPEND MOST OF THEIR LIFE IN THE OCEANS and migrate thousands of kilometres to breed. Only the females come ashore to lay their eggs on the beaches where they were born. Sea turtles are air breathers, but sleep under water for up to seven hours. Sea turtles are vital to a healthy marine habitat. Of the seven species, two are critically endangered, one endangered, three vulnerable and one is data deficient.

In Cape Verde, off West Africa, 20 years of beach patrols saw loggerhead turtle nest counts rise from 10,725 to 200,000.

Sea turtles have roamed all Earth's oceans, except its cold polar waters, for over 100 million years. But it only took the last 200 years of human activity to endanger all sea turtles. Millions of sea turtles were killed for their eggs, meat and shells, or accidentally caught in huge commercial fishing nets. Between 1844 and 1992, nine million hawksbill sea turtles were killed, mostly for their decorative shells. Further threats to these "cousins of the dinosaurs" are pollution, plastics, coastal development, disease, invasive species and climate change. Around 50,000 sea turtles drown annually when they get tangled in lost fishing gear.

The most effective way to improve sea turtle numbers is to protect the nests and eggs. In Cape Verde, off West Africa, over 175 kilometres of beach are patrolled by hundreds of rangers. Some rangers work with convicted poachers, teaching them to value the turtles. After 20 years, the Cape Verde loggerhead turtle nest count rose from 10,725 to 200,000.

Costa Rica's Tortuguero beach on the Caribbean coast was once a thriving market for the sale of sea turtle eggs, meat and shells. Today, it is a haven for green, giant leatherback, hawksbill and loggerhead turtles. The area's "Sea Turtles Forever" economy is based on protecting sea turtles for ecotourism.

Hawksbill sea turtles return

In 2007, the hawksbill was thought extinct in the eastern Pacific off Central America. This is because this turtle was given no protection. All eggs laid at Jiquilisco Bay, El Salvador, were destined for human consumption.

But for the last 15 years, it has a been a different story. Every egg is collected and taken to a centre to hatch. The hatchlings are then returned to the sea. To date, 250,000 hatchlings have been released to boost the number of critically endangered hawksbill sea turtles.

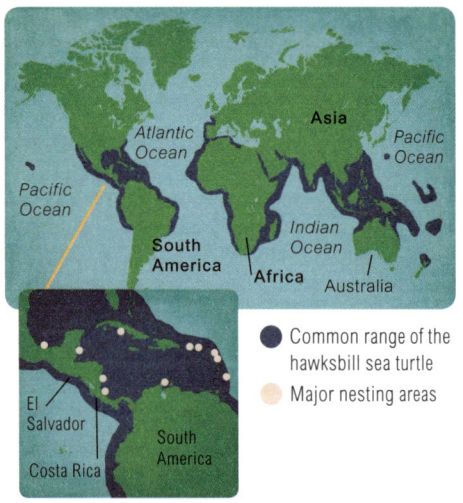

Numbers in the wild
6.5 million
(All species. Olive Ridley Project estimate, 2021)

Hawksbill sea turtles have a narrow pointed beak that allows them to prise sponges from narrow crevices in reefs. Among the smallest of the sea turtles, hawksbills are 89 centimetres long and weigh up to 68 kilos.

African elephants

Stopping the ivory trade, expanding sanctuaries, protecting habitats and encouraging human-elephant coexistence are just some of the steps needed to protect the world's largest land animal from extinction.

THERE ARE TWO SPECIES OF AFRICAN ELEPHANT. The savannah (bush) – the largest of all elephant species – is endangered, while the forest elephant is critically endangered. As a keystone species, elephants are crucial for the survival of every living thing within their wide range of habitats, from semi-deserts to tropical swamp forests.

Ivory from African elephants is mostly sold to Asia. The authorities there are using social media influencers to persuade their followers not to buy ivory products.

KaZa corridors

The Kavango-Zambezi Transfrontier Conservation Area (KaZa) was developed in Africa so that elephants and other animals could range freely across country borders. This vast park covers roughly 520,000 square kilometres of grasslands, wetlands, dry forests and scrublands.

Wildlife corridors now link 36 national parks, crossing between Botswana, Zambia, Angola, Namibia and Zimbabwe. Currently, there are 225,000 elephants in the KaZa. This is over half of Africa's total population of elephants.

- Distribution of savannah elephants
- Distribution of forest elephants

Numbers in the wild
415,000
(Both species. IUCN 2016)

Elephants have long been prized for their ivory tusks. Protecting elephants living within African reserves and in the wild from determined ivory poachers requires thousands of rangers and wardens patrolling on foot, and in vehicles and planes. Drones and infra-red cameras monitor herds. In some areas, elephants fitted with GPS collars are tracked day and night.

To give elephants the best chance of survival, herds are sometimes relocated from areas of danger to safer places. The biggest ever relocation happened in Malawi in 2016. It involved moving 500 elephants from two overpopulated reserves to a fenced and secure sanctuary over 322 km away. Here, the elephants had plentiful food and territory away from settlements.

In some places, humans and elephants have to live alongside each other. To reduce conflict between them, charities like Ecoexist in Botswana offer practical advice to local people. The charity encourages farmers to harvest their crops early to avoid seasonal raids by elephant herds. They also prevent land near established elephant corridors being leased for farming so elephants can move between their key habitats without conflict.

Standing nearly 4 metres to its shoulder and weighing nearly 6,000 kilos, savannah (bush) elephants can live 60 to 70 years in the wild. Sadly, thousands are killed every year.

Blue whales

The blue whale is a conservation success story. The combination of global "Save the Whale" campaigns and international bans on whaling has saved the world's largest mammal from extinction.

THE 30-METRE-LONG BLUE WHALE is at the top of the marine food chain. These whales are critical to the health of Earth's oceans, swimming through all of them except the Arctic. By eating 5,443 kilos of krill daily, they control the population of these small crustaceans. Whale faeces is a plentiful and nutrient-rich food for phytoplankton. In turn, these microscopic life forms photosynthesize, taking in carbon dioxide and giving out oxygen.

Some blue whale populations are increasing by 7 per cent a year.

Largest Marine Protected Area

In 2016, the Ross Sea off Antarctica was declared a Marine Protected Area (MPA). For a period of 35 years, commercial fishing is banned in this MPA which is more than ten times the size of England. This will protect the pristine ocean environment and whales, seals, penguins, seabirds, krill and many species of fish.

Preventing the overfishing of krill in the Ross Sea MPA means the blue whale's primary food source – krill – will be plentiful for its southern hemisphere summer migration.

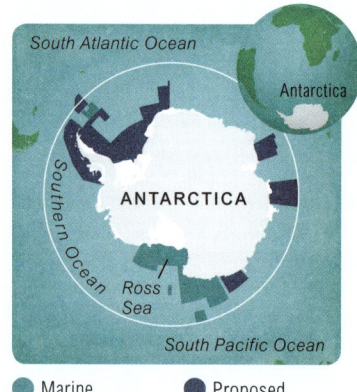

- Marine Protected Area
- Proposed protected area

Numbers in the wild
10,000–25,000
(All subspecies. IUCN 2018)

Over 110 years (1860s–1970s), the whaling industry almost made the largest animal known to have lived on the planet extinct. In just five years in the 1920s, half of all blue whales were killed. Their meat, bones and oils were used to manufacture lamp oil, candles, cosmetics, margarine, clothing and even umbrellas.

Public protests played a large part in making sure that whales have been protected since 1966. It is believed that no blue whales have been deliberately caught since 1978. In 1974, Project Jonah members rescued stranded whales from the beaches of New Zealand. Their work inspired other groups worldwide to do the same. The most famous of all conservation movements started in 1977, when Maris Sidenstecker and her daughter printed "Save the Whales" T-shirts. The money raised selling the T-shirts educated the public about the plight of blue whales.

In 2014, after four decades of conservation, blue whales in waters off the North American Pacific coast had recovered to almost pre-modern whaling levels. This population of blue whales is the only whale group to have done this. The IUCN Red List still ranks the blue whale as endangered, with climate change now a major concern.

Blue whales travel nearly 6,500 km in a year. They migrate – in groups or alone – to feed in cooler waters during the summer. They head to warmer waters near the equator to breed in winter.

Rhinoceros

There are five species of rhinoceros worldwide: three are critically endangered, and all need protection. Conservationists aim to save them all, even the Javan rhino, where only 18 mature animals are left.

THESE ONE-TONNE-PLUS ANIMALS CAN MEASURE 4 metres long and stand nearly 2 metres high. To keep such a large body fed, rhinoceros need to eat nearly 50 kilos of plant matter a day. The black and white rhinos graze the African savannah, the Indian (or greater one-horned) rhinos move between the grassland and forest, while Javan and Sumatran rhinos are generally found in rain and cloud forests. By simply eating, these grazers shape their habitats for the benefit of other animals and humans.

Less than 150 years ago, 500,000 rhinoceros lived in parts of Africa and Asia. Today, the worldwide mature population numbers around 16,000. Three species have become extinct in the last 25 years. Rhinoceros have been brought to the brink because of poaching and habitat loss. It is thought the only way to preserve the Javan rhino is to move some of its tiny population to new areas.

Trade in rhinoceros horn has been banned since 1977, but illegal poaching continues as the horn is used in traditional medicine. It is estimated that, in Africa, one rhinoceros is poached every 16 hours. It is not only rhinos living in the wild that are poached. Even those in fenced and patrolled reserves in Africa and Asia are vulnerable to poachers.

In addition to ranger patrols, two controversial techniques are used to reduce poaching. In one, rhinoceros are sedated and their horns removed. The horns regrow in 12–24 months. In the other, rangers infuse horns with a substance that makes people very ill if the ground horn is consumed. It is easy to spot a rhinoceros treated with this substance – its horns will be pink. The pink dye also makes the horn worthless as a decorative item.

Repopulating the Serengeti

As a result of poaching, the black rhinoceros population in Tanzania, Africa, fell to 123 in 2013. To increase population numbers in Tanzania's 31,000-square-kilometre Serengeti National Park, the Aspinall Foundation is transporting black rhinos from its wildlife parks in the UK to the Serengeti.

All rhinoceros are prepared for their relocation. They mix with animals, like giraffe, zebra, ostrich and wildebeest, that they will find on the Serengeti. They also learn to adapt to a fully-browsing diet. By 2019, nine rhinoceros had been reintroduced to the Serengeti with some producing calves.

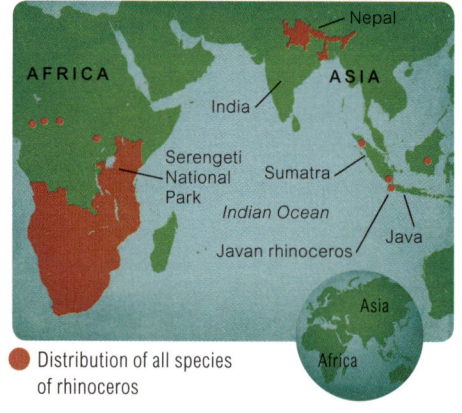

● Distribution of all species of rhinoceros

Numbers in the wild
15,470 mature individuals
(All species. IUCN 2018–2020)

These armoured giants have been on the planet for millions of years. They have only one predator – humans.

The black rhinoceros (illustrated here) and the white rhinoceros are not black or white. Both have grey skin. To tell them apart: the black one has a pointy upper lip while the white one has a squared lip.

Axolotls

Mexico's smiling amphibians have become critically endangered since their natural freshwater homes were polluted or drained. Ancient artificial islands and a concrete lake are helping their future.

LEGEND HAS IT THAT THE AXOLOTL IS THE AZTEC GOD of fire and lightning. The god disguised himself as an axolotl to avoid being sacrificed. This amphibian has the amazing ability to regenerate limbs and organs, and can live for up to ten years in the wild. It lives under water but breathes with lungs and gills, as well as being able to absorb oxygen through its skin. This fascinating creature is a popular pet but faces many problems in the wild.

There are several thousand axolotls in breeding centres. Scientists are looking for safe ways to reintroduce them into wild populations.

Recreating a safe habitat

Conservationists and farmers in Xochimilco, Mexico City, have been renewing and rebuilding ancient floating islands — they were originally designed by the Aztecs — in the shallow waterways. Known as "chinampas", they are made of layers of gravel, stones, native aquatic plants, logs and mud.

The chinampas provide land for growing crops, but also create safe, protected habitats in filtered unpolluted water where wild axolotls can breed and thrive. To date, around 5 km of axolotl chinampas refuges have been rebuilt.

- Current extent of lakes
- Chinampas

Numbers in the wild
Up to 1,000 mature individuals
(IUCN 2019)

Axolotls once thrived in the freshwater lakes and waterways around Mexico City, Mexico. Over the decades, increasing domestic, agricultural and industrial pollution have ruined their habitat. Then, in the 1970s, four of their lakes were drained for development to prevent flooding.

The axolotl had not just lost much of its watery home, but introduced fish, like tilapia, preyed on baby axolotls and competed for food. The waterways were then choked by the growth of introduced aquatic plants.

One of this 23-cm-long amphibian's remaining natural habitats is in Lake Xochimilco, Mexico City, but the tiny axolotl population is scattered over isolated areas. This limits the breeding opportunities that could increase axolotl numbers. However, naturalists have created breeding centres using axolotls from different populations, with the aim of reintroducing the axolotl to other suitable places.

Chapultepec is a park and zoo in Mexico City. In 2010, 26 adult axolotls were discovered in one of its cement-lined artificial boating lakes, along with three egg masses, and thankfully two of those egg masses hatched. Building on this natural success, Chapultepec is now home to the Axolotl Museum and Amphibian Conservation Center.

The feathery growths around the axolotl's head are its gills so it can breathe underwater. The axolotl also has lungs. It will swim to the surface for a gulp of air.

Lion tamarins

Lion tamarins are small monkeys. Once abundant in Brazil's forests, all lion tamarin species are endangered. When yellow fever infected golden lion tamarins in 2018 a vaccine had to be found – fast.

ALL FOUR SPECIES OF LION TAMARIN – golden, golden-headed, black, black-faced (superagui) – have a hairy mane. It is easy to tell one species from another by the colour of its mane and body fur. Lion tamarins are found only in forests along the Atlantic coast of Brazil, South America. These squirrel-sized New World monkeys spend most of their time in the forest canopy. They use all four limbs to hold on to branches.

Because of almost total habitat loss, the population of golden lion tamarins had shrunk to just 150 by the 1990s. To turn this around, New World monkeys needed access to more territory. Under pressure from conservation groups, the Brazilian government took action.

In 2007, they purchased two cattle farms that sat between a reserve and a large forest. The farms were reforested. This huge area was declared strictly protected. By 2014, the population of golden lion tamarins had grown to 3,200. Sadly, yellow fever, a disease spread by mosquitoes, reduced this to 2,500 in 2022. Work swiftly began to find a vaccine.

In another case, when an Atlantic coastal highway was widened, a section of forest was felled. This resulted in a small population of golden lion tamarins being isolated on one side of the highway. To reconnect the forests, an underpass and forested overpass were built in 2021. The primates can now safely cross the highway to mix and breed.

Some zoos have captive breeding programmes. Golden and golden-headed lion tamarin programmes have been successful, with similar ones now underway for the black lion tamarin. The aim is for these rare primates to be safely reintroduced to their wild habitat.

Vaccination programme

Yellow fever affects humans and non-human primates, like lion tamarins. The only way to prevent the virus wiping out the golden lion tamarin population was to create a vaccine for the primates, to make them immune.

This ground-breaking conservation programme, which started in 2022, requires capturing family groups in the wild and transporting them to a laboratory. Once injected with the vaccine, they are monitored before being returned to the wild. So far, around 300 golden lion tamarins have been vaccinated.

● Distribution of lion tamarin species

Numbers in the wild
9,250 mature individuals
(All species. IUCN 2000–2020)

The black lion tamarin was believed extinct for 65 years until rediscovered in 1970. This rare and vulnerable lion tamarin is now confined to one protected national park.

Golden lion tamarins are named after their red-gold colour. The largest of the lion tamarins, their bodies are about 25 cm long. A golden lion tamarin lives for about eight years in the wild.

Sea otters

Sea otters owe their survival to the world's first international wildlife treaty, agreed back in 1911. Today, they are still endangered, but they show how conservation work can really make a difference.

THESE POUCHED MAMMALS SPEND ALL THEIR LIVES IN WATER. They forage on the seabed for up to 14 hours a day. They dive for one to four minutes, then return to the surface to eat, usually floating on their backs. Sea otters are among the very few mammals that use tools. To crush the shells of crabs, abalones, clams and mussels, they bang them against a rock balanced on their chest. There are three subspecies of sea otter.

Oil pollution destroys the insulating quality of the sea otter's coat. An oil spill can kill thousands of sea otters.

Unlike seals, sea otters have no blubber. Instead, they have the densest fur of any animal to keep cold water away from their skin. This made their pelts highly valuable. Hunting reduced the population from hundreds of thousands in the late 1700s to 2,000 in 1905. It was US artist Henry Wood Elliott (1846–1930) who alerted the US government to the problem. He helped draft the North Pacific Fur Seal Convention, a treaty signed in 1911 by the USA, Japan, Russia, the UK and Canada. It banned open-water hunting of marine mammals for their fur. This treaty saved sea otters (as well as the northern fur seal) from extinction.

Sea otters are a keystone species. They control sea urchin populations that would otherwise devour whole kelp forests (kelp is a type of seaweed). Sea otters also eat mussels that would overrun rocky shorelines, leaving no room for other species. Despite this, problems such as oil spills and conflict with commercial fisheries mean sea otters today are listed as endangered by the IUCN. Many groups work to save them. From relocating otters to new areas to cleaning their fur after oil spills, they are finding new ways to save the "old man of the sea".

Monterey Bay Aquarium

Even though southern sea otter numbers are increasing, it remains an endangered population. This is why the Monterey Bay Aquarium, California, has a programme to rehabilitate abandoned sea otter pups.

Once rescued, each pup is bottle fed and groomed. Aquarium staff also help the pups learn to swim. A pup is then paired with a surrogate mother sea otter whose job it is to teach the pup how to dive and find food. When ready, the pup is released back into the wild.

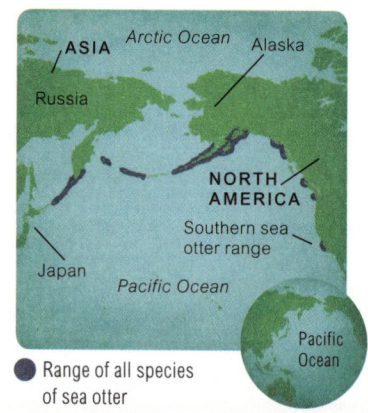

● Range of all species of sea otter

Numbers in the wild
128,902
(All subspecies. IUCN 2000–2018)

The three subspecies of sea otter are each found in distinct areas of the North Pacific Ocean. The southern (illustrated here) is found along the east coast of the USA. The northern sea otter inhabits waters off Alaska and British Columbia, while the Asian subspecies is located between far eastern Russia and northern Japan.

Conservation in Action

As we have seen in this book, there are many different ways that conservation organizations are working to protect wildlife. Here is a breakdown of the key areas that make up their activities.

1. Surveillance and research

As the work of the IUCN indicates, one of the first steps to conservation is keeping track of a species or a subspecies and its population, understanding what contributes towards its success or its decline. With this knowledge, conservationists can work towards building a population that can thrive in the wild, whilst keeping an ecosystem in balance.

2. Protecting ecosystems and biodiversity

No single species lives in isolation. The conservation of one species can benefit other species by protecting everything within the ecosystem, maintaining all the different forms of life in that area (biodiversity) and the natural food chain.

3. Educating local communities

It is important that conservationists accept that wildlife needs to live side-by-side with human populations. Involving local people in conservation helps ensure the long-term protection and survival of many species.

4. Promoting tourism and preserving local heritage

Emerging from working with local populations, conservationists recognize that their activities need to bring benefits to the places they are working in. A protected area with lots of wildlife encourages ecotourism and brings training, employment, and financial benefit to the community. Ecotourism should be done in a way that values and promotes local culture and history.

5. Encouraging sustainability

Sustainability means using and managing natural resources in a way that ensures they are not completely used up and can continue to be a resource for future generations. It means also understanding how the use of one natural resource affects another, and then working to ensure a balanced local and global environment for both humans and wildlife.

> "It is that range of biodiversity that we must care for – the whole thing – rather than just one or two stars."
>
> Sir David Attenborough (1926–), broadcaster, biologist, and natural historian

It is not just organizations that can help wildlife conservation. Our individual actions can make a difference too. These are just a few ideas.

1. Be a friend to nature

If you have access to a garden, think about how you can make it friendly to nature. Leave some of it wild and don't use toxic chemicals. Even if you don't have a garden, you can hang out feeders for birds or plant wildflowers in pots on your windowsill to attract bees and other insects.

2. Reduce, reuse, recycle

Reduce your energy use by turning off lights and make sure electrical devices are off and not on standby. Reduce water use by having showers rather than baths and not leaving taps running while brushing your teeth.

Think how you can reuse packaging or clothes rather than just throwing them away. Could you donate things you no longer want to charity? Remember: your junk could be another person's treasure. You can also enjoy buying pre-loved items from vintage and charity shops for yourself.

Learn about recycling in your local area – what can be recycled and what is rubbish – so that you can be sure your family is recycling as much as possible.

3. Be an educated consumer

There are many ways you can help the environment if you think before you buy. Don't buy highly-packaged groceries and avoid single-use plastics.

With your family, learn where your food comes from and try to buy foods that are grown locally. Research the guidance labels on what you buy generally, like sustainable, recyclable and humane, and use them to select your purchases.

Avoid products that harm animals and habitats. Examples of these products include micro-beads that can be found in some shampoos, soaps and household cleaning products; foods that have been grown, or use ingredients grown, using artificial pesticides; and products made with palm oil.

> **"There is a powerful force unleashed when young people resolve to make a change."**
> Dame Jane Morris Goodall (1934–), British primatologist and the world's foremost expert on chimpanzees

Conservation Word Bank

amphibians
A group of cold-blooded animals with backbones and smooth skin that live part of their lives in water and part on land.

apex predator
A meat-eating animal that hunts and eats many other animals but has no or few animals that prey on it. As a result, it is found at the top of a food chain.

birds
A group of warm-blooded animals with backbones, wings and feathers. Birds lay eggs and most fly.

breeding programme
A scheme to ensure the successful reproduction of an animal's young. This often takes place with captive animals held in zoos.

bushmeat
The meat from wild animals hunted by humans to eat.

climate change
The way the weather patterns (climates) vary over time. It is thought that human activity today, particularly the burning of fossil fuels such as oil and natural gas, is causing climate change through global warming.

corridor (natural/forest/wildlife)
A term used to describe a narrow area of land that is created to join up protected areas so animals can pass safely between them.

DNA
DNA is short for deoxyribonucleic acid. It is the substance found in all living cells that contain genes. Genes hold the inherited information that one generation of a life form passes to the next.

ecosystem
A geographical area made up of different living things and the physical elements they interact with, from animals and plants to soil and weather conditions.

ecotourism
Responsible tourism which is managed in a way that does not impact on nature and helps sustain local communities.

extinct
Describes a species that no longer has any living members. It has died out completely.

fish
A group of cold-blooded animals with backbones and fins that live under water breathing through gills.

genome
The complete set of genetic information held within a cell, made up of chains of DNA.

herbicide
A chemical used to kill plants, particularly those considered to be weeds.

insects
A group of animals without a backbone, who have a three-section body with a hard outer skeleton and six legs.

invasive species
A species of animal that invades or takes over a habitat it is not naturally found in. Invasive species are often introduced to a habitat, sometimes accidentally, through human activity.

keystone species
A species within an ecosystem that keeps it working well, connecting its various elements in a way that maintains a good balance between them. A keystone species is key to an ecosystem's survival.

mammals
A group of warm-blooded animals with a backbone and some fur or hair that breathe air. Female mammals produce milk to feed their young.

mangrove
A tree or shrub that grows in the salty water and swamps found in coastal areas. Most of its roots grow above the water.

marsupial
A type of mammal that gives birth to young at an early stage in their development and carries them in a pouch, feeding the developing young on its milk there.

migration
The regular journey made by some fish and other animals from one region to another. It is often linked to the seasons or breeding.

pesticides
Chemicals used to kill pests, including certain insects, that feed on and damage plants.

pet trade
The business of buying and selling animals for pets. It is often used in relation to exotic pets, which are pets from overseas, often rare and captured from the wild.

poaching
Hunting or catching wild animals illegally, usually for profit.

pollination
In flowering plants, the process of sexual reproduction where male pollen is transferred to the female part of a flower enabling the plant to make its seeds.

pollution
Damage caused to an ecosystem by the introduction of harmful substances, such as poisonous chemicals or waste.

raptor
A bird of prey of that kills or scavenges on other backboned animals for its food.

reptiles
A group of cold-blooded animals with a backbone and scales, that generally lay soft-shelled eggs, breathe air and live on land.

reserve
In conservation, an area of land or sea that is managed in a way that preserves it as an ecosystem. A reserve protects the different animals and plants that live there.

rewilding
In conservation, this can mean the reintroduction of an animal to its natural habitat. It can also refer to the process of allowing an area of land that has previously been farmed to return to its natural condition.

sanctuary
A place where wildlife is kept in safe, protected conditions.

savannah
An ecosystem found in tropical and sub-tropical regions which is mainly grassland with widely spaced trees and bushes.

scavenger
An animal that feeds on dead animals and plants or rubbish.

species
A group of living things that share the same characteristics and are able to reproduce together or share their genetic material with the next generation.

subspecies
A subdivision, or smaller group within a species, with minor physical or genetic differences from others in the species. A subspecies often develops when a part of a species becomes geographically isolated.

sustainable
Describes ways of managing resources to ensure their future supply without damaging the environment.

vaccine
A substance given to defend against a particular disease or infection by stimulating a living thing's natural ability (its immune system) to fight against it.

vulnerable
Under threat and in need of protection.

zoo
A place that keeps a collection of wild animals, usually in parks or gardens. Zoos study and conserve wild animals. They are generally open to the public.